An Introduction to
GRAND CANYON GEOLOGY

Michael Collier

TO MY FATHER,
who taught me
to look beyond the horizon.

In writing this book, I have bartered with and borrowed from many minds. Chuck Barnes and Arvid Johnson taught me their geology. Wesley Smith, years ago, told me about reading water and praying to the river. Arizona Raft Adventures and the Grand Canyon Natural History Association have, at one time or another, supplied valuable logistical support. Pat Miller, Tisha Cazel, and Margie Erhart critically sifted through the manuscript, each straining out a few more errors and uncertainties. Finally, Tim Priehs and Rose Houk have been the most incisive yet sensitive editors that a writer might wish to have. My thanks to you all.

FOREWORD

Probably few natural features of our earth have inspired more words — and more books — than the Grand Canyon. So why another? The answer must lie in the unique flavor of this author, who has done more than merely study the Canyon. He has let its shadows, its sounds, its whispering cascades, and its haunting peace become part of the inner person. This brief text is illuminated by that experience and that vision.

Michael Collier brings to this effort a list of impressive credentials. He received his bachelor's degree in geology from Northern Arizona University and master's in the same discipline from Stanford University. He has spent thousands of hours with the Grand Canyon as a geologist, photographer, writer, and veteran boatman on the Colorado River.

This book is an elegant melding of insights written for the layman who wishes a brief introduction to the geologic stories of the Grand Canyon region. On page one, it calls to us from nearly 13,000 feet atop Arizona's highest mountain, to share in the exhilaration of understanding Arizona's deepest chasm. As the last page is turned, the book deals with the sense of perspective the Grand Canyon forces on all of us. The book sings with enthusiasm and entrancing mystery.

An Introduction to Grand Canyon Geology begins by describing the Colorado Plateau, that vast frame for its most conspicuous ornament, the Grand Canyon. Chapters two and three discuss the origins of, first, the upper flat-lying rock layers, and then the dark and tangled rock at the bottom of the Canyon's Inner Gorge. Chapter four deals with the awesome earth forces which have buckled, broken, and hoisted the Colorado Plateau and the Grand Canyon region. Having examined the various rock layers and their patterns of folding and faulting, only one thing is missing—the Canyon itself. Chapter five describes the carving of the Grand Canyon by the Colorado River.

Savor this *Introduction to Grand Canyon Geology* slowly and refer to it often. It will bring you closer to understanding some of the basic questions which revolve around this magnificent natural feature. What after all does the Canyon mean? It may simply be a brilliant scene to be wondered at from the many rim overlooks and catalogued among one's treasured memories. Or it may be a reminder of the power of the human mind.

The stories in the Canyon's walls were mute until questioning people gave them form. It is an awesome paradox that the same geologic knowledge that makes us seem so insignificant in time comes from the only creature in 46 million centuries that denied his own insignificance with an intellect groping to understand it all.

A part of that heritage is in your hands. It was minds like yours that saw limestones and read epitaphs.

Charles W. Barnes
Flagstaff, Arizona
February, 1980

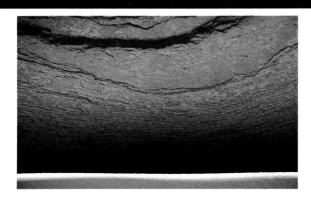

An Introduction to Grand Canyon Geology
Copyright 1980 by the Grand Canyon Natural History Association
Library of Congress Number 80-67863
Editorial: T.J. Priehs, Rose Houk
Production: T.J. Priehs, John C. O'Brien, Don McQuiston, Debra McQuiston, Robert Petersen
Design: McQuiston & Daughter
Illustration: John Dawson
All the photographs appearing in *An Introduction to Grand Canyon Geology* were taken by Michael Collier.
Cover photographs by Tom Bean
ISBN 0-938216-04-X

CONTENTS

Foreword, by Charles W. Barnes iii

1. A View from the Top 1
   An Introduction to the Colorado Plateau
2. Water Rocks, Wind Rocks 5
   Sedimentary Rocks
3. My Deepest Thoughts on the Inner Gorge 13
   Metamorphic and Igneous Rocks
4. Bumps and Grinds 21
   Structural History
5. Erosion and Landforms 29
   Erosional History
6. A View from the Bottom 39
   An Epilogue
Synopsis: A Brief Two-Billion-Year History
   of the Grand Canyon 41
Bibliography 41

A VIEW FROM THE TOP

Mountains are a wonderful distraction. I drove through Flagstaff last summer, toward an appointment in the Bradshaw Mountains one hundred miles to the south. I didn't want to be late. The San Francisco Peaks, at first ahead and then off to my right, watched as I drove around their bases. Agassiz watched. Humphreys Peak watched.

I'd not climbed Humphreys Peak in two years. I had stayed the night, that time before, watching the mountain's sunset shadow race eastward to meet the moonrise. Magic had been afoot. I remembered all this and, still driving, set my resistance like a south-bound sail. No, I would not climb the peaks this morning. There were places to go, people to see.

Four hours later I was sitting on top of Humphreys, breathing hard. It would have been easier had I succeeded in talking the chairlift operator into letting me ride up the first two thousand feet for free. No deal, she smiled. I had to walk.

The wind swirled in from the southwest. I'd brought along my recorder. I played the first few notes of "Four Strong Winds." The wind carried the music in long streaming banners to the northeast. I saw a C-sharp fall on the ruins at Wupatki, and a B-flat float over Grand Falls. One plaintive G drifted all the way to Navajo Mountain.

The song faded and I silently drank in the land laid out before me. Gray Mountain, the Hopi Buttes, and the Henry's. Painted Desert and the Marble Platform. The Vermilion and Echo Cliffs. Shadow Mountain and Sunset Crater. Enchanting names. They all belong to a land called the Colorado Plateau—130,000 square miles of visual poetry. A fantasy land. It is my home.

Geologists have carved the United States into about a dozen geologic provinces, each based on

an area's generalized similarities. The Colorado Plateau is one such province, occupying about half of Utah and progressively smaller portions of Arizona, New Mexico, and Colorado. It is outlined by volcanoes and lava flows along its southern margin in Arizona and New Mexico, by the Rockies in Colorado, and the fault-lined deserts of the Great Basin in Utah.

What similarities do these 130,000 square miles share? For the most part, the Plateau is covered by relatively flat-lying sedimentary rock, ranging from 50 to 550 million years old. Sedimentary? A rock composed of fragments of other rocks or derivatives of the sea, deposited by wind and water. But more about that later. This veneer of sedimentary rock ranges in thickness from four miles in northern Utah's Uinta Basin, to nonexistent where it has been eroded away at the bottom of Arizona's Grand Canyon.

It is not particularly unusual to find so much sedimentary rock piled up in one place, but it is unusual to find it lifted a mile or more above the sea level at which it was deposited—without being seriously folded, torn, or crumpled in the process. Indeed, the Colorado Plateau was elevated 5000 to as much as 13,000 feet in the recent past (geologically speaking), all without being unduly squashed.

To be sure, some wrinkles and bumps did occur during this uplift. To the north, there is the Waterpocket Fold—a sharp 3000-foot wrinkle in Utah's Capitol Reef National Park. To the west, the world tumbles 2000 feet over the brink of the Hurricane fault. But these folds and faults are the exception. The dominant impression is one of long low land swells, running forty miles to the

south and 400 miles to the north, beyond the horizon and into my imagination.

This is a dry land. While its peaks and higher mesas are carpeted in pine and fir, the lower surface is given to grass, sage, and barren rock. Dry washes are scratched across its skin. Sand dunes drift up the face of the Adeiie'echii and Echo Cliffs out beyond the waterless bed of the Little Colorado River.

Water in a dry land. A precious, precocious commodity. There have been times when I was hiking here cotton-mouthed with dry heat, yet listening nervously for the rumble of a sudden flood. When water does come to this land, it can arrive by the canyon-full. Flash floods erode through bare soft rock, to fashion growing canyons and receding cliffs. These unique patterns of erosion support the region's claim to being a separate geologic province. The Plateau is, above all, a land of canyons.

The Colorado Plateau is roughly pyramid-shaped. Its eastern margin is drained by the Little Colorado, San Juan, Dolores, Colorado, and Green rivers. They converge toward the apex of the pyramid, that is, toward the south and west. Along the way, these rivers carve some of the most elegant canyons in the world—Labyrinth and Stillwater, Westwater, Cataract and Glen canyons. Their walls of Navajo and Cedar Mesa sandstone rise and fall like the tentative chords of an orchestra that is rehearsing before a symphony. This canyon music swells as the Colorado River crosses into Arizona and flows past Lees Ferry into Marble Canyon. With the arrival of the Little Colorado from eastern Arizona, the apex is reached, the symphony begins. The Colorado River, having assimilated its major tributaries, cuts a final most majestic canyon. The Grand Canyon.

The Canyon (and in the Southwest, THE Canyon carries the same singularity of meaning that THE City evokes in New York) is only fifty miles north of the San Francisco Peaks. I sit a mile higher than the Canyon's rim, straining unsuccessfully to see the Colorado River that is still another mile below that rim.

The Canyon's North Rim stands a thousand feet above the opposing South Rim, allowing me to study the northern walls a third of the way down, about to the level of the red Supai beds. I can't really see much detail—it *is* fifty miles away. The 9000-foot Kaibab Plateau stands out nicely hovering over the Canyon's northern edge. I find it easier to follow the upper geologic formations as they bend through the great East Kaibab Monocline and dive into the abyss of Marble Canyon. It is not so easy to find the monocline's counterparts on the western side of the Plateau—

the Big Springs, Noble, and Crazy Jug faults. They are lost in the distance.

The Grand Canyon is quintessential canyon country. The bones of the land here lay bare. White ribs of Kaibab Limestone and Coconino Sandstone shine through the distance's softening haze. The dynamic balance between the uplift of mountains and the downcutting of erosion has momentarily been frozen in the walls of the Canyon. Geology as cinema: running slowly enough that we can examine time frame by frame.

But the Colorado River and its tributaries must one day erase the Kaibab Plateau, 9000 feet high and forty miles long. Indeed, the river, flowing through the Grand Canyon, must carry the entire Colorado Plateau, grain by grain, year by year, out to sea.

Geology is the study of revolution in the largest sense of the word. Atop the San Francisco Peaks, I can feel the earth slowly turning. Gray Mountain buckles up through buried layers of the past; the rains of time slowly wear it down. In land such as this, geologic explanations underlie every observation. In land such as this, rocks have voices and the canyons sing.

Summary

The Colorado Plateau covers 130,000 square miles in the Four Corners states. It is a geologic province characterized by sedimentary rocks lifted a mile or more above sea level, and by the ubiquitous canyon-carving processes of erosion in an arid land. The Colorado River drains most of this area, creating its most awesome canyon just before leaving the Plateau—the Grand Canyon.

Royal Arch Creek, a tributary to Marble Canyon.

Bright Angel Creek is a mess. The rocks that line its walls are twisted and torn. Rock layers appear and disappear in flagrant disregard for the more obvious guidelines of stratigraphic geology.

Years ago, when I was first getting to know the Canyon, I spent the better part of a week trying to unravel the puzzling Bright Angel fault that parallels its namesake creek. I scrambled into side canyons and up cliff faces, mapping out each rock type that is exposed along the canyon. It would certainly have been easier (and probably more accurate) to simply read Jim Sears' paper on the fault or John Maxson's earlier work on the canyon. I had no illusions of discovering new revelations that would change geologists' basic understanding of the Bright Angel scenario. Mostly I was looking for an excuse to poke around in the Canyon.

I had been mapping for two days when I ran into a problem. Near Clement Powell Butte, the Tapeats Sandstone abruptly ends against an outcrop of Shinumo Quartzite. Sandstones aren't supposed to end abruptly. It's against the rules. They are laid down, usually by seas, in great continuous sheets that can cover thousands of square miles. Traced laterally, one expects a layer of sandstone to grow thinner and pinch out, or perhaps to slowly change its composition to a siltstone or conglomerate. But there was no gentle transition or thinning of the Tapeats Sandstone here. There was just an abrupt change to the Shinumo Quartzite, a rock twice as old as the Tapeats. No faults were close enough to explain the change. I was buffaloed.

Late the second afternoon, a light clicked on. That abrupt boundary, I speculated, must be the last vestige of an island in the sea which deposited the Tapeats Sandstone. A fossil island,

WATER ROCKS, WIND ROCKS

A

B

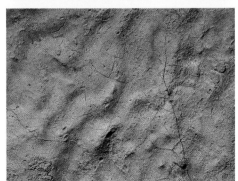

C

D

E

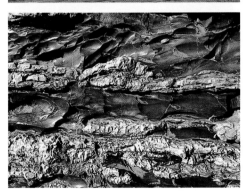

Sedimentary rocks contain many clues that help explain their origins: A) Brachipods, fossilized in the Redwall Limestone, once lived at the bottom of a warm, clear sea. B) Ripple marks in the Bass Limestone were formed in water currents more than a billion years ago. C) Worms burrowed through the mud and silt that was to become the Bright Angel Shale. D) Crossbedding in the Coconino Sandstone resulted from the migration of wind-blown sand dunes. E) Alternating bands of dark chert within the Redwall Limestone possibly indicate fluctuating silicon concentrations in ocean waters 330 million years ago.

half a billion years old! (In retrospect, it was high time for me to arrive at this notion. Geologic literature about the Canyon is crawling with references to these islands.) This island, made up of the resistant Shinumo Quartzite, would have presented a bold line of cliffs to the Cambrian sea that was engulfing it. The sea nibbled at these cliffs even as it slowly buried the island beneath its sediments.

The outcrop in question was two miles away; running over to it, I shinnied out on a geologic limb and predicted that I would find evidence of boulders crashing from the island into the sea.

Geologists must often satisfy themselves with abstract solutions to equally abstract problems. Evidence is painstakingly compiled to imply the presence of some ancient adjacent landmass or shoreline. Past environments are reconstructed grain by grain. Geology is more often a cerebral exercise than many people may realize. But that afternoon in Bright Angel Canyon I was treated to an exhilaratingly tangible display of one small nest of secrets locked inside the earth.

I reached the wall near Clement Powell Butte.

Shinumo Quartzite—a billion years old—soared in cliffs over Tapeats Sandstone that was only half its age. Angular blocks of quartzite, some the size of small houses, were imbedded in the sandstone at the base of the island. Layers of sandstone were crushed and splattered by the impact of the Shinumo boulders. Some of the smaller boulders had rolled out into the sea, but most were right at the base of the ancient cliff. I listened for the echo of rocks falling. But all I could hear was the sea crashing against that cliff, half a billion years ago.

The Shinumo and Tapeats formations are part of the sedimentary veneer that covers the Colorado Plateau. All told, twenty-one sedimentary formations are exposed in the Grand Canyon. Stacked one above another, these strata give us a view of the earth that stretches back a billion years.

A billion years.

What can that possibly mean? The earth, we are told, is 4.6 billion years old. The Bass Limestone, at the bottom of this sedimentary stack, is a quarter as old as the earth itself. This immenseness eludes me. Time shrinks into a nebulous concept and disappears behind numbers on a page.

Hiking down the Hermit Trail, I once found fossil corals in the Redwall Limestone. I looked into the tubes that had housed each animal and imagined them beneath the sea, straining sea water for food to earn their living. A tedious existence, I decided. Envisioning myself as a coral, I became bored of straining sea water.

Suddenly it occurred to me that an instant which took place in the Mississippian period, 330 million years ago, had snapped abruptly into focus, up from that elusive background of unfo-

Flutes on a limestone boulder, sculpted by the Colorado River.

cused time. Hiking farther down the Canyon, I would rub elbows with trilobites and worms that lived 500 million years ago. I imagined wriggling through silt that was to become the Bright Angel Shale. Focus again. Pinpointing these moments among millions of years, I began to appreciate the otherwise incomprehensible vastness of time.

The sedimentary rocks of the Grand Canyon hold a step-by-step record of much of the earth's history. To be sure, the record is neither complete nor easily deciphered. But it waits here for those who wish to understand it. John Wesley Powell, first to run the Colorado River through Grand Canyon, put it this way in 1875: "One might imagine that this was intended for the library of the gods; and so it was. The shelves are not for books, but form the stoney leaves of one great book. He who would read the language of the universe may dig out letters here and there, and with them spell words, and read, in a slow and imperfect way, but still so as to understand a little, the story of creation."

Sedimentary rocks are deposited by water and wind. Geologists, examining the Canyon layer by layer, have reconstructed the sequence of environments responsible for these layers; warm seas, desert dunes, quiet lagoons. The processes of sedimentation are not haphazard. An environment which deposits a sandstone one time will not capriciously deposit a limestone some other time. Trusting this, we can examine the walls of the Grand Canyon and read its story. The sedimentary chapters of this story cover two distinct periods.

From either rim, the most striking feature of the Canyon's walls are the continuous horizontal bands of rock. These are the uppermost (and thus the youngest) sedimentary strata, forming the top 4000 feet of the Canyon. They are representative of the great Paleozoic era, spanning ages from 250 to 550 million years. From top to bottom, we see the advance and retreat of no less than seven seas, one Sahara-like desert, and at least a few lagoons.

With a little patience it is possible to differentiate not only the obvious white Coconino Sandstone and the massive cliffs of the Redwall Limestone, but also the more subtly defined Muav and Toroweap formations. With a lot of practice and a little luck, you may even distinguish the four newly defined formations within the Supai Group, an ability which still eludes many geologists possessing plenty of practice but not enough luck.

Beneath these horizontal Paleozoic strata lie the tilted Younger Pre-Cambrian rocks, inclusively (and impressively) referred to as the *Grand*

7

Canyon Supergroup. This rock band contains, among seven others, the aforementioned Shinumo Quartzite and the Bass Limestone. Their ages are all in the neighborhood of a billion years, give or take a couple hundred million. Fossils are rare in rocks this old, but evidence of algae is found in the Kwagunt Formation (800 million years old) and possibly in the Bass Limestone (1.2 billion years old).

Sometime after its creation but before that of the overlying Paleozoic strata, the Grand Canyon Supergroup was broken into a series of tilted north-south trending mountain ranges along faults such as the Bright Angel. Some blocks were lifted; others were dropped. Erosion beveled off all but the lowest of these blocks. As a result, we find only isolated pockets of the Grand Canyon Supergroup scattered throughout the Canyon. Indeed, the upper half, called the Chuar Group, is found only in a great valley below Point Imperial on the North Rim. In a large number of locations within the central and eastern Grand Canyon, we find remnants of Shinumo Quartzite rising as ancient resistant islands within the Tapeats Sandstone and its overlying Bright Angel Shale.

When the flat-lying Tapeats Sandstone was laid down above these tilted rocks, an angular relationship was established that John Wesley Powell in 1869 would name the Great Unconformity.

The Grand Canyon plays tricks on our eyes, on our minds. Like the woodcuts of M.C. Escher where one man will see birds and another will see fish—in the Canyon one person may see rocks while another hears the whispered passage of time. Products to one, rocks may imply process to another. There is a sequence of Paleozoic rocks in

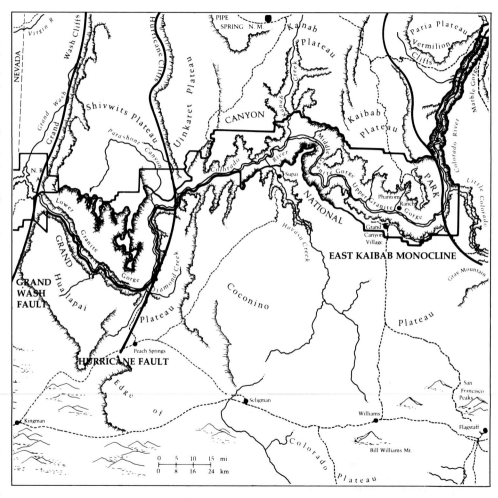

the Grand Canyon which beautifully illustrates this dichotomy.

Many folks, staring into the Canyon without much to say, have noticed the following sequence of Paleozoic rocks just above the Inner Gorge: tan cliffs of the Tapeats Sandstone beneath the blackbrush-covered terraces of the Bright Angel Shale, beneath the gray-green rubbly cliffs of the Muav Limestone.

Sandstone, shale, limestone. The boundaries between these layers are not sharp; instead they are gradational, one blending into the next. Trudging up from the river, you would have a chance to observe that the particles composing these rocks become smaller as you climb. Sandstone is made of larger grains than shale; shale of larger grains than limestone. Let's figure out why.

All three rocks are sedimentary, laid down by an ocean. This sea was moving onto the land, advancing from west to east. Let's arbitrarily pick a point for reference—say, Indian Gardens, below Grand Canyon Village—and follow what happens to it. Remember, we must go back to a time when there was no Grand Canyon. The rocks that make up its upper 4000 feet weren't here. All that existed was the sea to the west, and low flat land to the east.

As the sea first covers this pre-pre-historic Indian Gardens, crashing waves and shoreline currents churn the water. Rivers from the east dutifully deliver their loads of sediment—sand and silt and mud. But only the heaviest grains of sand can fall to the bottom. The mud and silt are too light and stay suspended in the murky water. They are carried farther west, out to sea. Only deposits

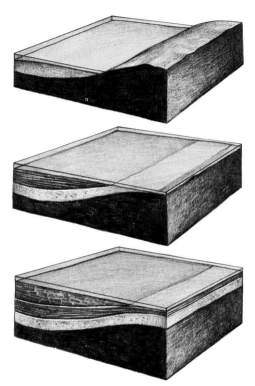

The sea (water level indicated in blue) is moving onto the land (transgressing). At first only heavy grains of sand accumulate on the bottom. As the sea moves farther onto the land lighter particles of clay and silt accumulate. Eventually the shoreline has moved so far away that carbonate materials are the predominant deposits.

of sand are laid down.

The sea presses eastward. The waters at Indian Gardens have grown calmer because the shoreline—with its breaking waves and longshore currents—has moved east. Progressively smaller particles are able to float down and be deposited on the bottom. In time, deposition of sand is overshadowed by the accumulation of silt and mud. But the sea still marches east. Eventually the shoreline is so far away that very little river-borne silt and sand is able to reach our submerged outpost at what is now Indian Gardens. In place of the silt and sand, calcium carbonate is deposited. Had we dug down into the ocean floor at this point, we would have first encountered calcium carbonate, then shale and silt, and finally, lowest of all, sand.

Oceans teem with life. In fact, this ocean, at the beginning of what is called the Cambrian period, 550 million years ago, witnessed an unprecedented, never-to-be-repeated explosion of life forms. Trilobites, molluscs, and brachiopods were scuttling all over the place, flashing fat bank rolls and talking out the sides of their mouths. It was a boom chapter in the annals of Life. These critters lived, died, and sank in staggering numbers. Where conditions were favorable, their hard shells built up, layer upon layer, and in time became rocks.

Sea water contains calcium carbonate, from which animals build their shells. Under the right conditions, this compound can precipitate directly out of sea water and be deposited on the ocean floor. These two sources of calcium carbonate—shells and precipitants—either singly or together, form a type of rock called limestone.

9

In modern oceans about a yard of limestone is deposited every 7500 years. But sandstones in a shallow ocean can typically accumulate a yard's thickness every 1500 years; shales, a yard every 3000 years. So, while calcium carbonate may be deposited along with river-borne sediments, its presence will probably be masked by the more quickly accumulating sands and silts. Thus, limestones are most likely to be deposited farther from shore than sandstone or shale.

So there it is—the Cambrian sequence presented in all of its petrologic glory. Sand, silt, calcium carbonate. Sandstone, shale, limestone. The Tapeats, Bright Angel, Muav. Take a little time and look at these rocks again. Do you see only rocks, or do you also hear the stories, do you feel the persistent undercurrent of the earth's history tugging at your mind? Do you see product or process? If geology merely catalogued dead rocks, I for one would have abandoned my interests in it long ago. But it is more than that. Geology is an anthology of stories, the best in the world.

Summary

Sediments—sand, mud, and organic debris—are laid down by oceans, rivers, and wind, to become sedimentary rocks. There are twenty-one sedimentary formations in the Grand Canyon, stacked one on top of another. The lower eight formations, of Younger Pre-Cambrian age, were tilted sometime after their deposition. The line between these and the upper thirteen, those of Paleozoic age, is called the Great Unconformity. These twenty-one sedimentary rocks were laid down in a wide range of environments—quiet oceans, shallow seas, swamps, and deserts.

Kaibab Limestone, 300 feet. Cliff and ledge former.
Middle Permian (250 million years old)
Composed of light tan sandy limestone. The sea which deposited the underlying Toroweap Formation quickly returned to deposit the Kaibab Limestone. Fossil evidence, including molluscs, crinoids, and brachiopods, indicates a warm, shallow marine environment of deposition.

Toroweap Formation, 200 feet. Ledge and cliff former.
Middle Permian (260 million years old)
Composed of tan sandstones and limestones, with interlayered beds of gypsum. The Toroweap records the advance and retreat of yet another sea from the west. From east to west, the Toroweap Formation varies from a predominant sandstone to a limestone. Brachiopods, molluscs, corals, and bryozoans are found within this formation.

Coconino Sandstone, 50-300 feet. Cliff former.
Early Permian (270 million years old)
Composed of tan and cream-colored sand, deposited in great dunes now preserved as crossbedded sandstone. The Coconino Sandstone records a great Sahara-like desert that covered northern Arizona in early Permian time. Many small reptiles left tracks now fossilized in the Coconino. Curiously, no fossils of the animals themselves have ever been found.

Hermit Shale, 300 feet. Slope former.
Early Permian (280 million years old)
Composed of bright red siltstones that are easily eroded to a low slope. Originally combined with the underlying Supai Group, the Hermit was defined as a separate formation by Levi Noble in 1922. Like the Supai, the Hermit was deposited in swamps and lagoons.

Supai Group, 600-700 feet. Ledge and cliff former.
Pennsylvanian/Permian (300 million years old)
Composed of red siltstones and an upper red-tan sandstone. Eddie McKee, of the U.S. Geological Survey, in 1975 redefined the old Supai Formation to include the Watahomigi, Manakacha, and Wescogame Formations and the Esplanade Sandstone. Great crossbeds are the trademark of the Supai group. Mud cracks and plant fossils found in the Supai indicate a depositional environment that was low and swampy.

Redwall Limestone, 400-650 feet. Consistent cliff former.
Early-Middle Mississippian (330 million years old)
Composed of fine-grained gray limestone with bands of chert at some horizons. The Redwall contains abundant fossil evidence of crinoids, brachiopods, bryozoans, and other fauna typical of a warm, shallow, clear-water ocean. This limestone was deposited under an unbroken rain of carbonate debris, resulting in the massively bedded cliffs that we see in the Canyon today. Rainwater flowing down the overlying Supai and Hermit redbeds has stained this formation.

Temple Butte Limestone, 100-1000 feet. Cliff former.
Devonian (~370 million years old)
Composed of red-purple thinly-bedded dolomite and a fine-grained gray thickly-bedded dolomite. The Temple Butte forms channel-shaped deposits along the upper surface of the Muav Limestone, especially conspicuous in Marble Canyon. To the west, the Temple Butte becomes a thick cliff former, reaching 1000 feet near the Grand Wash Cliffs. Fossil remains are limited to fish teeth that were not destroyed during the post-depositional conversion of the Temple Butte into a dolomite.

Muav Limestone, 150-800 feet. Cliff and ledge former.
Middle Cambrian (530 million years old)
Composed of mottled gray limestone with green micaceous siltstones. Deposited offshore in a shallow sea, the Muav often felt the effects of storms which ripped up the bottom. The Muav is the uppermost evidence of the sea

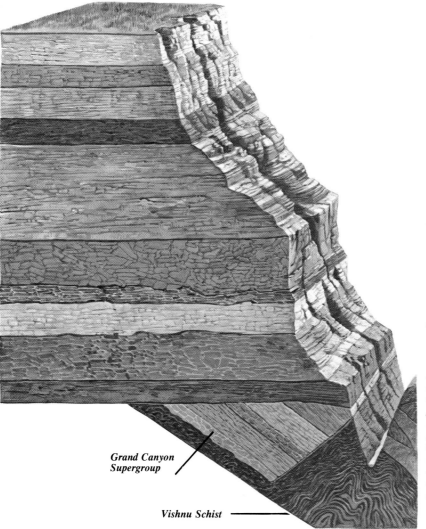

which advanced into this area during the Cambrian. Dolomite deposits within and above the Muav reflect the regression of this sea back to the west.

Bright Angel Shale, 200-450 feet. Consistent slope former.
Early-Middle Cambrian (540 million years old)
Composed of shaly green mudstones, with some fine-grained sandstones near its contact with the underlying Tapeats Sandstone. This shale formation contains many fossil remains of trilobites, brachiopods, and worms. Like the Tapeats, the Bright Angel Shale becomes more fine-grained toward its top, where it too gradationally changes into the rock that is above—the Muav Limestone. The Bright Angel was deposited in the quiet waters, farther from shore, of the same sea responsible for the Tapeats Sandstone.

Tapeats Sandstone, 100-300 feet. Cliff former.
Early Cambrian (550 million years old)
Composed of medium to coarse sand grains, in colors ranging from cream to deep red brown. Diagonal crossbeds are the preserved surfaces of ancient coastal sand dunes. The Tapeats was laid down beneath the shallow and turbulent coastal waters of a sea advancing from the west.

Grand Canyon Supergroup, 15,000 feet
Younger Pre-Cambrian (1.2 billion to 800 million years ago)
A tremendous thickness of sedimentary rock and interbedded lavas covered the Grand Canyon region before the beginning of the Paleozoic era. These rocks, all formations within the Grand Canyon Supergroup, ranged from soft shales to resistant sandstones. A period of mountain-building broke them into blocks and tilted them 10 to 15 degrees. Erosion then removed all but the lowest blocks. Flat-lying rocks were subsequently deposited above these tilted rocks during the early Cambrian period, creating the angular relationship called the Great Unconformity.

Up to this point I have managed to shirk one of my responsibilities as a communicating geologist. I haven't informed you that the earth's surface is composed of three kinds of rocks. This is akin to a chemist not mentioning that atoms are made of electrons and nuclei. Or a used car salesman casually forgetting to mention that the '57 Hummer he wants to sell has no engine. So, without further ado. . . .

The earth's surface is composed of three kinds of rocks. Sedimentary. Igneous. And metamorphic. Sedimentary rocks, already introduced, are the layered deposits of water and wind. Igneous rocks are fire-born, heated within the earth until they melt and rise again to the surface. Metamorphic rocks are the baked and squashed products of an earth-moving unrest that troubles the land beneath us.

Each of these rocks can be reshuffled to become one of the others or itself again. Any of these three rock types can be eroded and washed seaward to eventually be deposited as a sedimentary rock. Rocks can be buried beneath the earth's surface, heated to melting, and then surge back up as igneous volcanoes. Or they can be trapped within the vise of mountain-building pressures, to be squeezed in the presence of heat until, without melting, their constituent minerals become unstable and reform as metamorphic minerals, as metamorphic rocks. Collectively these comings-and-goings of rock types are known as the rock cycle. A sort of geologic mandala.

The Grand Canyon, as we have already seen, is well endowed with sedimentary rocks. Igneous and metamorphic rocks are also important here. These two types are both the oldest and the youngest rocks in the Canyon. They are the lowest and the highest. Beneath the Paleozoic and

MY DEEPEST THOUGHTS ON THE INNER GORGE

13

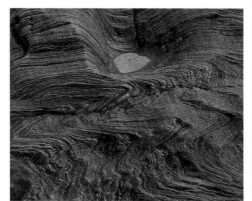

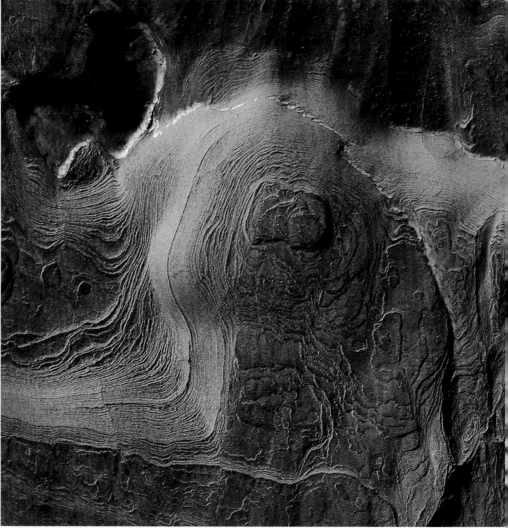

Younger Pre-Cambrian strata lies the metamorphic-walled Inner Gorge. Viewed from Grand Canyon Village, the Gorge is a dark abyss at the Canyon's bottom, swallowing both the river and our imaginations. The river struggles into view only in isolated spots. Trails, visible from the rim, disappear into the Gorge and seem not to return. What's down there?

The Vishnu Schist is down there. Of all the rocks in the world, the Vishnu must be my favorite. One morning years ago, rowing through the Gorge near Clear Creek, I sidled up to an enticing outcrop of schist and scrambled onto the rock. The Vishnu here was brimful with garnets, a metamorphic mineral. They were hard and rounded, the size of grape seeds. Resisting the river's urgent call to be worn down and carried away, the garnets stood in relief above the surrounding rock surface. Each garnet had created a tiny eddy in the current; softer minerals, hiding in these shadows, had been protected from erosion.

Midnight rock, showered with garnets. It was as if each garnet were a small comet with its tail sweeping out behind. Sky and rock. Sometimes I think that nature enjoys mimicking herself.

The Vishnu Schist bears an impressive pedigree. It dimly records the most ancient episodes of

Left, *the Vishnu Schist, a metamorphic rock, underlies all other rocks in the Grand Canyon. Two billion years ago, the Vishnu existed as shales and basalts. But given heat, pressure, and a great deal of time, individual atoms within the rock reorganized themselves into new minerals to become the components of a schist. Above right,* pink granite is intruded into the darker schist.

geologic history legible in the walls of the Grand Canyon. Shales and volcanic rocks accumulated here on an earth only half so old as it is today. Two billion years ago, perhaps more. We can no longer see these rocks. Instead, by examining the minute textures of the Vishnu Schist, cataloguing its mineralogic content and calculating its chemical composition, we can infer the prior existence of shales and volcanic rocks. These rocks, in turn, infer an origin in a particular environment—low lands and shallow coastal waters, punctuated by volcanoes and lava flows. The landscape would have been hopelessly stark, plantless and lifeless. The sky, carbon-rich and oxygen-poor, probably was not the blue we see today. Blue-green algae and simple bacteria drifted in the ocean, biding their time.

How, then, did these sedimentary and volcanic rocks manage to transform themselves into a new and totally different rock, into the Vishnu Schist? To invoke the vernacular of Plate Tectonics—a geologic master plan that is all the rage these days—things must have become hot and heavy. To illustrate with a contemporary example . . . take a continent, say Asia, in your right hand, and another, say India, in your left hand, and bang them together. What do you get? Mt. Everest and its Himalayan cousins. At the core of such a mountain range, temperatures and pressures would reach levels capable of metamorphosing solid rock.

Our original sedimentary and igneous rocks, buried beneath thousands of feet of additional sediments, may have been forced further underground by a mountain-building episode that occurred 1.7 billion years ago. Buried so far inside the earth, they would have become hot, something on the order of 900 degrees Fahrenheit, and subject to pressures in excess of 100,000 pounds per square inch. Things had indeed become hot and heavy. Minerals, stable in their former environments, would have become unstable. Aluminum, iron, silicon, and other ions could begin their restless reorganization toward a new existence.

The stage is now set for the crystal-by-crystal creation of the Vishnu Schist. Minerals like garnet, along with muscovite, staurolite, and hornblende begin to appear with the gentle mystery of stars at dusk. Creation is an act which escapes the resolution of our microscopes, our sciences. It is so pervasive, so fundamental, that we must ultimately accept it as a mystery. Geology is a science great enough to have left room for its miracles.

Metamorphic crystals nucleate in random orientations. But some crystals like muscovite are

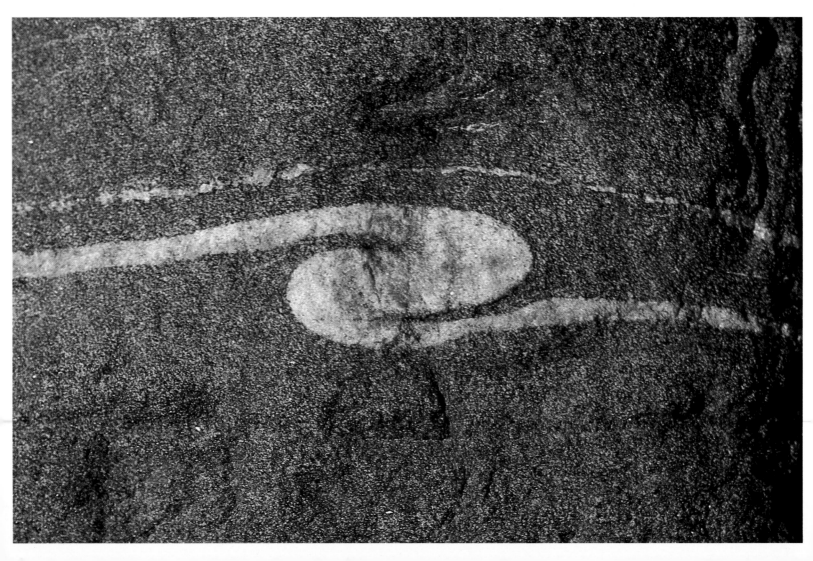

flat; others like hornblende are elongate. If the mountain-building compression which began this metamorphic process still existed, it would have been easier for these minerals to grow in some directions than others. Muscovite and hornblende crystals oriented in a plane perpendicular to the direction of compresson would tend to grow faster than those in less favorable orientations. Trillions upon trillions of these crystals, all aligned in roughly the same direction, give rise to a texture in schist called foliation.

The Vishnu is permeated with fine striations, usually running up and down outcrops of the rock. These are the planes of foliation. Looking into the Inner Gorge even from the Canyon's rims, you can see the effects of this alignment. Notice the vertical slashes in the schist, caused by erosion following the patterns of foliation. Since the foliation patterns are vertical, the direction of the mountain-building compression responsible for metamorphism of the schist must have been approximately horizontal.

Still looking into the Inner Gorge, you may be able pick out vertical bands of light-colored rock, contrasting with the dark Vishnu Schist. This is the Zoroaster Granite. Igneous rocks, which include granite, are the third of geology's three primary rock types. Originally molten, the Zoroaster intruded along, and occasionally across, the foliation planes of the schist. Slowly cooling far

Left, *quartz instrusion within the Vishnu Schist.* Above right, *Columnar joints within this basalt wall formed as lava cooled, shrank, and cracked.*

below the earth's surface, its constituent minerals had sufficient time to grow into large crystals.

Igneous rocks, by definition, must once have been molten before freezing into their present form. Liquid rock is considerably more mobile than solid rock. It can readily travel *from* someplace *to* another place. Where the Zoroaster Granite went seems obvious. It went into cracks in the Vishnu Schist. But where did it come from?

The Zoroaster probably rose from a pocket of silicon-rich magma (molten rock) lurking beneath the Vishnu Schist. Since the material was molten, atoms and ions were relatively free to migrate within it. Since the magma was injected into the schist deep underground, it would have been sufficiently insulated to cool very slowly. Ion mobility, coupled with this extended cooling period, allowed crystals to become very large. As crystal grew up against crystal, an interlocking mesh of granitic rock was formed.

When did all this take place? The Vishnu Schist had been metamorphosed about 1.7 billion years ago. The granite, necessarily younger than the schist into which it was intruded, was emplaced 1.5 or 1.6 billion years ago. These dates may be taken either on faith or on the strength of radiometric dating, neither of which I will attempt to justify in this brief text.

Molten rock, it was noted earlier, travels *from* some place *to* another place. This simple expository statement begs further examination. What if magma doesn't stop travelling inside the earth to become an *intrusive* rock, but rises to the surface to become an *extrusive* rock? To answer this, we take you live and direct to the western Grand Canyon. The Canyon's best examples of extrusive rocks—landforms such as lava flows and cinder cones—are found here along Toroweap Valley and the Hurricane Cliffs. If you have been taken west via raft down the Colorado River, you are probably becoming a trifle nervous. For days you have been hearing rumors about That-Riffle-At-Mile-179.6.

Lava Falls. This rapid offers the true test of a geologist's loyalty to his science. Will he, in the face of such an improbably violent rapid, be able to notice the cascades of black lava that once poured into the Grand Canyon and froze onto its walls? Will he examine this rock with its small olivine crystals, glassy matrix, and columnar joints, and conclude that it must have cooled quickly after flowing into the Canyon? Will he notice the cinder cones on the Canyon rims above? Will he remember that this rock is only a million years old, making it the youngest of all the Canyon's rocks?

Probably not.

Summary

Geologists divide all rocks into one of three classes—sedimentary, metamorphic, and igneous. Metamorphic rocks are represented in the Canyon by the Vishnu Schist. The Vishnu, once an assemblage of sedimentary and volcanic rocks, was metamorphosed 1.7 billion years ago at the heart of a great mountain range. Igneous rocks, once molten, are divided into two groups—intrusive, such as the Zoroaster Granite, and extrusive, such as the western Canyon lava flows—depending on whether they cool beneath or above the earth's surface.

Right, the Vishnu Schist is composed of minerals that are aligned along planes of foliation, determined by the direction of compression during the schist's metamorphism. The schist was subsequently folded in some locales. Facing page, The Colorado River *now polishes the Vishnu Schist at the bottom of the Inner Gorge.*

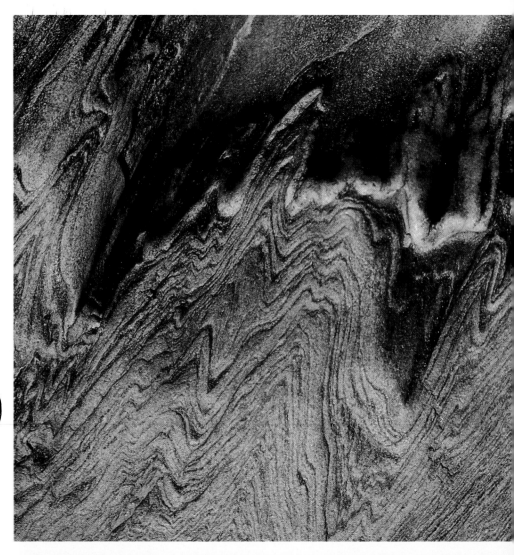

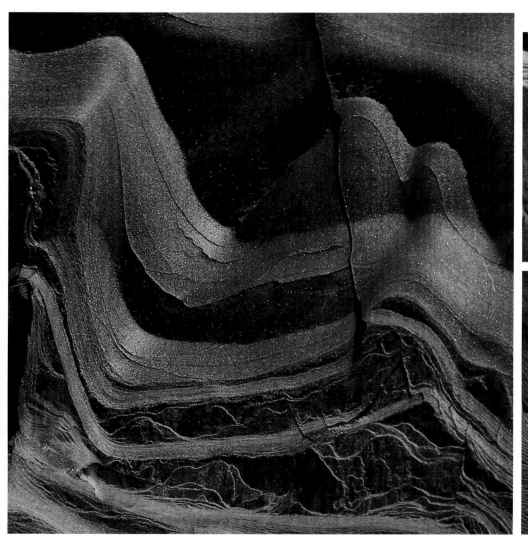

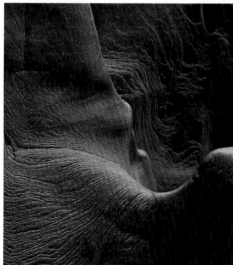

Geology is a young science. Its most fundamental tenets were suggested little more than a century or two ago. In 1869, as John Wesley Powell first explored the bottom of the Grand Canyon, serious scholars were still debating the relationship of Noah's Flood to fossils found in mountainsides. James Ussher, Archbishop of Armagh, announced during the mid-seventeenth century that the world was created on October 26, 4004 B.C.

About one thousand years ago (or one-sixth of the earth's age, according to the Archbishop), an Anasazi Indian laid a ram's horn on a ledge along the back wall of a small stone room, beneath a fold in the Grand Canyon's Tapeats Sandstone.

One thousand years later: my graduate research involved study of a system of small tight folds in the Muav Limestone near Havasu Canyon. I had chosen this problem for a thesis because it could only be reached by rafting down the Colorado River. Boatmen will do anything for one more trip—even graduate geology research. Before reaching the study area near Havasu, I encountered a beautiful fold in the Tapeats Sandstone. It was vaguely similar to the folds I would study further down river. I stopped to investigate, hoping that it might shed light on my folds in the Muav. It didn't.

But I did find the foundations of a small stone room. I was puzzled. There was no sheltering overhang to make this site more desirable than any other. Why would an Anasazi build a room here? Then I found the ram's horn, so old that I first mistook it for a piece of juniper wood. This room and its ram's horn were nestled in that beautiful fold within the Tapeats. Perhaps this site had not been chosen by accident; the Anasazi, I speculated, had been paying homage to the same earth forces that I was attempting to understand.

BUMPS AND GRINDS

21

Geology may be a young science, but mankind has felt the power of this earth for a long, long time.

Looking across the river, I followed the same fold up through the Bright Angel, Muav, and Redwall formations. A small side canyon, hewn from the red Supai siltstones, suggested a vertical continuation that I was unable to observe directly. Each rock, with its characteristic patterns of mechanical behavior, had reacted differently to the forces responsible for the fold. The Tapeats was kinked into a tight 'S' curve; its many layers of sandstone had been able to slide one over another while bending. The massive Redwall Limestone, on the other hand, had behaved almost monolithically, as it bent through one gentle but ponderous curve.

These rocks had initially been deposited in flat layers. What forces could have stretched or compressed them to the point that they would behave like pliable plastics and bend into the shapes that we now see exposed in the Canyon walls?

Throughout the Paleozoic (550-230 million years ago) and Mesozoic (230-70 million years ago), the land that was to become the Colorado Plateau remained relatively peaceful. It rose a little here, fell a little there; seas came and seas went—the same old sedimentary story. But a twenty-million-year episode of rock-buckling and mountain-building—called the Laramide Orogeny—brought this tranquility to an abrupt end.

Beginning at the close of the Mesozoic era, the Laramide Orogeny was destined to change the face of western America. Out of this turmoil grew the Rocky Mountains from Canada to New

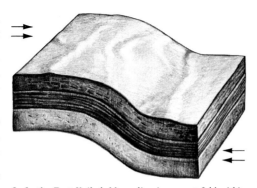

Left, *the East Kaibab Monocline is a great fold within the earth's crust. Rock layers to the left are 3000 feet higher than their counterparts to the right.* Above, *rock layers, originally flat, are compressed and buckle into a fold.*

Mexico, the Uintas of Utah, the Wind Rivers of Wyoming. And the Colorado Plateau.

Laramide wrinkles on the Plateau vary in size from the small fold in the Tapeats Sandstone where I found the ram's horn to the great East Kaibab Monocline, which defines the eastern margin of the Kaibab Plateau. Some of these features stretch only a few miles and then disappear, while others can be traced lengthwise for hundreds of miles. Each is unique. And though it is possible to dwell on the differences of each feature, it may be more fruitful to examine their similarities.

Squeeze this book from left and right, and you will generate a fold that runs up and down. Squeeze the Colorado Plateau from east and west, and you will generate folds that run north and south. The majority of Laramide features on the Plateau are folds that trend approximately north-south. It is logical to assume, then, that the Laramide Orogeny was an east-west compressive event.

The North Rim, at 8000 and 9000 feet above sea level, is improbably beautiful in early October. The aspens are a riot of yellow and gold. From Point Imperial, at the eastern edge of the Kaibab Plateau, you can look 3000 feet down to the Painted Desert on the far side of the Canyon. Aspens, fir, and spruce surround you at Point Imperial, but only eight miles away the Painted Desert is sparingly clothed in scrub brush and sage. Point Imperial and the nearby Painted Desert are floored by the same rock—the Kaibab Limestone. The difference in its elevation is here due to 3000 feet of uplift across the East Kaibab Monocline.

Kaibab is the Piute Indian word for "mountain lying down." It is their name for the high plateau that extends north from the Canyon, fifty miles toward Utah. Being a blunt Anglo-Saxon modern geologist, I would most likely have compared the Kaibab Plateau with a half-watermelon-lying-down. (Oh well. Edward Abbey saw fit to call the desert sun an infernal-flaming-plasmic-meatball-in-the-sky. *Sic transit gloria linguae*.)

Many Laramide folds in the vicinity of the Grand Canyon take the form of monoclines. Curiously, almost all of the Plateau's monoclines are east-facing; that is, rocks to the west of the folds are most frequently higher than the same rocks on the east side. Various mechanical and stratigraphic explanations for this situation have been suggested, but none is conclusive. It is satisfying to know that geology is riddled with unanswered questions of this sort. A science with all the answers and no more questions atrophies into mere technology.

Monoclines found in the walls of the Grand Canyon display another intriguing pattern: many are directly above ancient faults in the Vishnu Schist. These faults, breaks in the earth's crust, were active in Pre-Cambrian time. Their orientation suggests origin in a tensional environment. Stretched apart, one block was free to slide down past another block. But when Laramide compression began 65 million years ago, the blocks were squeezed back up in the opposite direction. Thus they are called *reverse faults*. Why would monoclines be located directly above these reverse faults? Geologists have grappled with this problem for some time without producing a single comprehensive explanation that is thoroughly satisfying.

The eastern edge of the Kaibab Plateau is outlined by the East Kaibab Monocline. But what structure defines the western edge of the Kaibab Plateau? To the west we see no smooth continuous fold. Instead, a series of north-south breaks exists—the Crazy Jug, Noble, and Big Springs faults—which drop rocks on their western flanks back down to the 6000-foot elevation more typical of the Colorado Plateau.

The geometry of these faults suggests a tensional rather than compressive environment. Tensional and compression, by definition, cannot exist simultaneously in the same north-south direction in the same rocks. The faults to the west of the Kaibab Plateau must, then, have been formed at some time other than Laramide.

The Colorado Plateau, following all the com-

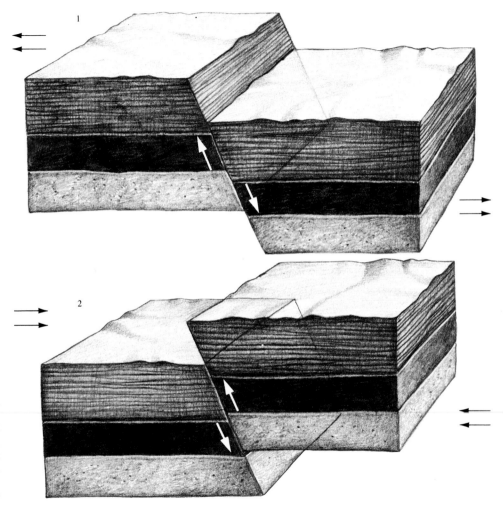

pressive hubbub of the Laramide Orogeny, became a nice quiet neighborhood. (Or, for any geologists remaining in the audience . . . "the Plateau returned to tectonic quiescence.") Mastodons and primeval rhinoceros trundled across the Plateau, methodically pursuing their evolutionary task of making little mastodons and rhinos. A pastoral scene, indeed.

This uneventful period droned on for 25 million years, to be terminated by a second mountain-building episode. Beginning 20 million years ago, during Miocene time, the American Southwest was stretched apart and torn into a series of basins and mountain ranges.

The Basin and Range Province is a geologic entity comparable to the Colorado Plateau. The

tension that began in Miocene times, though expressed more dramatically in the Basin and Range Province, also left its mark on the Plateau. The Crazy Jug, Noble, and Big Springs faults on the west side of the Kaibab Plateau were formed at this time, along with larger faults 50 to 75 miles farther west. These large faults—the Hurricane and Grand Wash—define the boundary between the Colorado Plateau and the Basin and Range Province.

The Hurricane fault, measured north of Hurricane, Utah, offsets rocks by at least 4000 feet. The Grand Wash fault may have slipped as much as 16,000 feet near its intersection with the Colorado River. In addition to being torn along these faults, the Colorado Plateau was uplifted to its

Left, *a reverse fault. This type of fracture pattern often requires a two step process: 1) A normal fault develops as a block of rock is stretched and then torn apart, 2) tension in the rock is then replaced by compression. Adjacent blocks move relative to one another, but in a direction reverse that of the original movement.*

Right, *a normal fault. A block of the earth's crust is stretched by tensional forces until the block fractures. One half of the block then drops down relative to the other half.* Above right, *Monument Fold has contorted the beds of Tapeats Sandstone near Elves Chasm.*

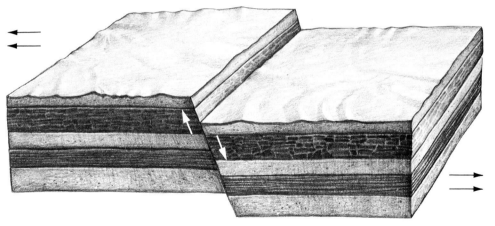

present mile-high altitude during this period of tensional unrest.

Faults and folds are the meat and potatoes of structural geology. With the close of the Miocene, the tectonic fireworks of the Grand Canyon region begin to sputter and subside. The Canyon's major faults and folds are in place, though some of the established faults were to remain active until almost recent times. To wit: the Toroweap and Hurricane faults have been active recently enough to slice through basalts that froze into place less than a million years ago. But no new major faults or folds have been formed at the Grand Canyon since Miocene time. A structural geologist might now be tempted to close up shop and go home, ignoring the Canyon's more recent history. After all, he explains, the only thing left for nature to do at this point is to finish eroding the Grand Canyon into the shape we see today.

There remains, however, a number of problems potentially fascinating to the structural geologist. Great blocks, isolated by erosion from the main Canyon walls, have in some cases broken away and are slowly subsiding *en masse* into the Canyon. The faults along which the blocks have broken are termed *high-angle gravity faults*. Vishnu and Jupiter temples in the eastern Grand Canyon are dissected by these faults.

Surprise Valley, between Tapeats and Deer creeks, offers another challenge to the structural geologist. Here, four square miles of rock slid into the Canyon temporarily plugging up the river sometime in the geologically recent past.

Beaver Falls in Havasu Canyon.

It appears that the shales of the Bright Angel Formation—perhaps after lubrication by ground water or by water backed up behind lavas flowing into the Canyon further downstream—were unable to support the weight of the formations in the walls above. Consequently, much to the surprise of the unsuspecting geologist, entire sequences of the lower Paleozoic have slid down and are found considerably lower than where they belong.

A third example of recent rock movement is found at river level in the Bright Angel and Muav formations near Havasu Creek. These rocks have been neatly bent into a series of small tight folds that precisely follow the sinuous river for twenty-five miles. Gravity—a quantifiable force—provided the impetus for this fold system. Since the amount of bending within the folds can be measured, this fold system offers the structural geologist an ideal opportunity to whip up some nasty looking mathematical equations that relate stress to strain in a natural laboratory.

Each of these phenomena—the gravity faults, Surprise Valley, and the river-oriented folds—suggests that the Canyon is still adjusting to the shock of its having been cut so deep, so fast. In the next chapter, armed as we now are with a knowledge of its rocks and underlying structures, we will investigate the cutting of the Grand Canyon by the Colorado River.

Summary

Solid rock can be torn up and kicked around in any number of ways, each method producing its own geologic result. In the Grand Canyon we see Laramide compression and Miocene tension creating various faults and folds, uplifts and downdrops. Compression can form folds such as the East Kaibab Monocline. Tension can form faults like the Hurricane and Toroweap faults. All of this structural activity occurs on scales ranging from the small river-oriented folds near Havasu Canyon, to the uplift of the entire Colorado Plateau.

27

The Colorado Plateau is a land of secrets. Its surface rolls along horizon-bound beneath a blue sky, juniper covered, sage colored. This visual serenity belies the depths of so many canyons that are hidden within its limestone and sandstone. Not far from Lees Ferry, at the head of Marble Canyon, lies a wildly beautiful chasm. Its walls pinch and swell, swirling to the music of crossbedded sandstone. Their shapes mimic the rush of the summer flash floods which have carved them from the Navajo Sandstone. At the bottom are places where you can't see the sky. The walls are that narrow. On top, the long-legged among us could step over the canyon and not break our stride. The Plateau is indeed a land of secrets.

Having, pages ago, taken the liberty of speculating on the boredom of a Mississippian coral and on the slitherings of a Cambrian worm, I feel justified in imagining the plight of one García López de Cárdenas. His boss, Francisco Vásquez de Coronado, in the course of searching for the elusive Seven Cities of Cíbola in 1540, had heard rumors of a sizeable river to the west of his route. He dispatched Cárdenas to investigate.

Cárdenas was led by Hopi guides across the Painted Desert toward what is now the vicinity of Desert View. Walking westward through juniper and pinyon pine, the explorer, without warning, suddenly found himself staring into the Grand Canyon. He was the first European ever to do so. The "sizeable river" a mile below looked a mere trickle, six feet wide by Cárdenas' first estimate. To stand at the brink of this unknown Canyon was to nibble at the very ear of oblivion. I am quite sure that Cárdenas walked a little ways back into the forest and sat down for a moment. A silly confused grin covered his otherwise hardened explorer's face. The Plateau never really prepares

EROSION AND LANDFORMS

us for the first sight of its secrets.

What process lies at the heart of the Plateau's secrets? What force in nature could carve canyons so whimsically as they seem to be carved on the Colorado Plateau?

It's the water.

Water, flowing downhill, knows two tricks. It can carry fragments of soil and rock in its current. And, with these particles as teeth, water can gnaw away new particles from old rocks. These interrelated actions, called transport and erosion, are two sides of the same geologic coin. Climate, geologically recent uplift, and the flat-lying sedimentary rocks have all combined with the erosive process of flowing water to form the scrimshaw canyons of the Plateau.

The process of erosion is slow. Major Powell would be hard pressed to see many changes in the walls of the Grand Canyon since passing this way a century ago. But this process is all that nature has to carve her canyons. And if the Grand Canyon be used as evidence, it must be enough.

The Colorado Plateau is an arid land. Much of it fits the geographer's definition of a desert, receiving ten inches of rain or less in a given year. This precipitation is concentrated in two seasons—summer monsoons and winter snows. Runoff tends to collect rapidly in washes and arroyos and rush down rivers as floods, explosively powerful floods.

In December of 1966, the North Rim of the Grand Canyon received fourteen inches of rain in thirty-six hours. Forty-foot walls of water, moving at upwards of eighty miles an hour, scoured Bright Angel and Crystal canyons. Debris carried by Crystal Creek clogged the course of the Col-

Above, *Deer Creek Falls*. Right, *Thunder Spring*.

orado, creating overnight one of the most nerve-wracking rapids in the river's repertoire.

Historically, the river, swollen with the meltwaters of winter, would rise every spring, flooding the floor of the Canyon. The debris brought in by side canyons during the previous year would be swept downriver. Millions of tons of sediment were daily washed along the river bed.

But in 1963, Glen Canyon Dam began impounding the waters of the Colorado River, just upstream from Lees Ferry. The prodigious sediment loads now silently settle to the bottom of Lake Powell. The dam, with a maximum release capability of 30,000 cubic feet per second, can't begin to simulate the spring floods that have been measured at 300,000 cubic feet per second. Young rapids—like Crystal and the gaping hole at Granite Park which formed in 1978—are now left untempered by the old spring floods.

In time (60 years say the optimists; 300 years say the pessimists) Lake Powell, filled with sediment, will be a marsh and the river will flow over the top of the Glen Canyon Dam. The erosive power of water in a 700-foot fall will flick the dam from the river's path like a twig. The redwater floods of spring will return. I take solace in this: the river, even in the face of our most earnest tinkerings, will not be stopped.

The Colorado River, then, has carved *down* through at least a mile of solid rock. But this action is limited to its riverbed. The Grand Canyon is thirteen miles wide in places. What causes the *outward* growth of the Canyon?

The answer, in part, lies in the web of tributary creeks that carve back into the walls of the Canyon. These creeks, falling 5000 and 6000 feet in a

matter of miles, carry a tremendous potential for erosion and transport. But again, this potential is concentrated at the creek bottoms.

The outward growth of the Canyon walls is due primarily to gravity. In a process called weathering, moisture seeps into fractures in rocks and corrodes the cement that binds them together. Repeated freezing and thawing will pry the rocks open. Weathered rocks, once balanced above cliffs, are loosened and fall to the canyon floors. The creek below will soon carry these fragments into the river. And the river will grind them down and wash them out to sea.

Weathering, erosion, transport.

In previous chapters, we examined the formation of sedimentary, igneous, and metamorphic rocks that comprise the walls of the Grand Canyon. And we saw how these rocks were twisted and torn into structures such as faults and folds. These processes, for the most part, were at work long before the Colorado River began to carve its Grand Canyon. Already in place, the rocks, faulted and folded, would directly influence the Canyon's eventual patterns of erosion. The study of these influences and their resultant landforms is contained within the science of geomorphology, little sister to geology itself.

We can use geomorphic lines of thought to investigate three of the Grand Canyon's fundamental features. First let's consider the differences between the North and South rims. The most pronounced differences between the two rims involve side canyon lengths and the availability of water. Side canyons that cut into the North Rim are typically longer and have more free-flowing water than their South Rim counterparts.

The Kaibab Plateau has been irreverently alluded to as a half-watermelon-lying-down. The river flows south, parallel to the length of this watermelon (read: Kaibab Plateau) and then turns west to cut across its southern nose. Formations to the north of this cut (read: Grand Canyon) are higher than the same formations to the south. Ground water on the north side tends to flow south, toward the cut; ground water on the south also tends to flow south, but away from the cut.

The walls of the North Rim are consequently blessed with a wealth of large springs—Thunder River, Roaring Springs, Cheyava Falls, to name a few. The waters of these springs have followed the layers of rock on the North Rim that dip south into the Grand Canyon. Meanwhile, the South Rim limps along on the likes of Santa Maria Springs below Hermits Rest and Indian Gardens. These springs are laboring even to water the mules and hikers along the Bright Angel and Hermit trails. The North Rim, 1000 to 2000 feet higher than the South, also receives considerably more precipitation. These factors—direction of ground water movement and the differences in surface water quantities—serve to carve longer and larger side canyons on the North Rim.

A second geomorphic puzzle considers the vertical profile of the Canyon's walls. Looking into the Canyon, we see steep cliffs at some levels and variously pitched ledges and slopes at other levels. The rim of the Canyon is formed by a cream-colored cliff of Kaibab Limestone. In a more rainy climate, this limestone would have been softened into an easily eroded slope. But on the arid Colorado Plateau, the Kaibab Limestone characteristically forms a sharp cliff. A thousand feet below the rim, the bright red Hermit Shale forms its typical slope. The Hermit is friable and soft, easily eroded.

Throughout the Grand Canyon, the great Redwall Limestone exists as a fortress-like barrier. Harvey Butchart, that most intrepid of Canyon hikers, logged only one hundred routes through this barrier in all of the Grand Canyon. The Redwall is massive limestone, seamlessly deposited beneath a quiet and continuous rain of calcium carbonate. There are few natural breaks within it which erosion can now transform into slopes and ledges.

Three formations below, the Bright Angel Shale forms its continuous, level platform within the lower Canyon. The Grand Canyon's longest hiking trail, the Tonto, stretches some seventy miles along this platform. Again, the soft shales, easily eroded, are the influence behind the shape of this particular landform.

The Canyon's vertical profile, then, is a relatively simple problem in geomorphology. Somewhat more complex is a study of the horizontal profile.

The Colorado River is best described in three segments. The first segment flows south-by-southwest, out of Utah and down through Marble Canyon. The second segment—the Grand Canyon—begins at the confluence of the Little Colorado River and ends at the Grand Wash Cliffs on the western edge of the Colorado Plateau. The third segment of the river flows across basin and range landscape, through Lake Mead and then south to the Gulf of California.

The Colorado River's path through Marble Canyon is easy enough to understand: it is flowing

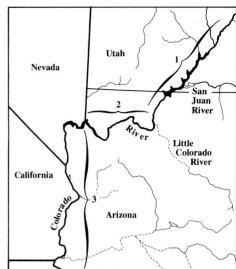

Left, *Chuar Creek carving out Lava Canyon.* Above, *three segments of the river figure most importantly in the history of the Grand Canyon. The first segment includes Marble Canyon and the now-drowned Glen Canyon. The second contains the Grand Canyon, and the third includes the low country from the Grand Wash Cliffs to the Gulf of California.*

along the lower flank of the East Kaibab Monocline. The monocline has existed long enough (65 million years) to have influenced the course of the river in its younger, more impressionable days. Most geologists believe that the river has occupied its Marble Canyon site for a considerable length of time—since at least the Oligocene epoch, some 30 million years ago.

This explanation works nicely until the river, in its second segment just below the Little Colorado River, abruptly turns west and runs headlong into the 9000-foot high Kaibab Plateau. Why would a sensible river like the Colorado suddenly take a notion to flow up over (and then cut down through) the Kaibab Plateau? This is tantamount to asking "why is the Grand Canyon here?" Geologists tackled this question the minute they first laid eyes on the Canyon in the nineteenth century. And I am delighted to report that they haven't quit arguing about it since.

Powell advanced the notion that the Kaibab Plateau rose against an already established Colorado River. The river would have cut through the Plateau like a stationary saw cuts through a rising log. This eloquent explanation, however, doesn't mesh with facts found just west of the Grand Wash Cliffs. There, the Muddy Creek Formation contains rocks that had to be deposited before the Colorado River flowed along its present course through the Grand Wash Cliffs. The top of the Muddy Creek Formation is radiometrically dated at about six million years. Thus a through-flowing Colorado River must be younger than six million years old and could not (as Powell suggested) have been established before the Kaibab Plateau rose during the Laramide Orogeny.

Charles Hunt in the 1950s began to play with the idea that the Colorado River avoided the problems of the Muddy Creek Formation by seeping beneath the Grand Wash Cliffs in a system of underground tunnels that would have put the entire Central Arizona Project to shame.

But Hunt's proposal foundered, in part, because the third segment of the Colorado River

Facing page, *the Redwall Limestone, easily eroded in water, is dissected into crevasses along the Colorado River in Marble Canyon.* Left, *the Colorado River, brown with silt from its tributaries, is wearing away the schist and granite of Bedrock Rapids.*

(where it now defines Arizona's border with California and Nevada) could not have existed so early. The Gulf of California had opened for business four to five million years ago. The lower Colorado River had begun draining into it then, gradually extending its domain northward by carving an increasingly longer valley—a process called headward erosion.

In 1964 Eddie McKee convened a round table at the Museum of Northern Arizona in Flagstaff to sort out all this confusion. McKee and company concluded that the ancestral Colorado River had indeed flowed down through Marble Canyon, but it then turned up the Little Colorado River to eventually spill into the Gulf of Mexico. Meanwhile, small tributary creeks on either side of the Kaibab Plateau were supposed to be carving toward one another by the processes of headward erosion. When they met, the westerly flowing stream captured the drainage of the easterly flowing stream. At this point, the Colorado River began to flow to the west as we know it today, through the Kaibab Plateau.

This hypothesis was hampered by a small hitch, however. No one ever found the ancestral river bed of the Colorado where it was supposed to flow east and south across Arizona, New Mexico, and Texas. Back to the drawing boards.

Ideas, like canyons, seem to grow by headward erosion. They slowly carve their way into uncharted territory. Encountering one another, one idea captures the flow of the other. An explanation of the Colorado River's history that has gained favor with a lot of geologists was proposed by Ivo Lucchitta in the mid-70s. Hunt had tried to drain the ancestral Colorado River to the south-

west, and then McKee had tried draining it to the southeast. Lucchitta's theory has the river running around the southern end of the Kaibab Plateau (as we see it today), but then draining north into Utah. The river, according to Lucchitta, had encountered the Kaibab Plateau after the Plateau was formed (65 million years ago) but before erosion of the surrounding countryside had isolated it as a topographic barrier. Large basins, isolated from the sea, existed in southern Utah during Tertiary times and could have received the ancestral Colorado. When the Gulf of California opened four to five million years ago, the lower Colorado River grew by headward erosion toward the north-flowing ancestral Colorado, eventually capturing its waters.

There are problems with this interpretation. The eastern Grand Canyon, with its steep walls and vigorous rapids, has a shaggy youthful appearance that belies the more dignified age that Lucchitta would assign it. Unfortunately, the Utah basins have not yet been thoroughly investigated so that we are not sure that they received the ancestral Colorado River. Perhaps more disconcerting, the western Grand Canyon would have had to cut to within a few hundred feet of its present depth in something like three million years—a blistering rate of erosion. Of course there will be refinements to Lucchitta's explanation of the Canyon's history. And of course geologists will continue to argue about it until the river itself dries up.

There is a children's story by Kenneth Grahame that I delight in reading to my river passengers during the summer. I usually let eight or nine days

Left, *confluence of the Little Colorado River. The Colorado River, green and cold, contrasts with its warm brown tributary. The two rivers flow side by side for half a mile before mixing.* Above, *Saddle Canyon.*

of the two-week trip go by before reading it. By then, the Canyon has had a chance to slowly reveal its history to these people. They have heard all the formation names, have learned about faults and folds and horizontal profiles of the Canyon. They have had a chance to listen to rippling waters and singing walls.

The story is about a water rat, a mole, and their riverbank friends. "The Mole," begins Grahame, "was bewitched, entranced, fascinated. By the side of the river he trotted as one trots, when very small, by the side of a man who holds one spellbound by exciting stories; and when tired at last, he sat on the bank, while the river chattered on to him, a babbling procession of the best stories in the world, sent from the heart of the earth to be told at last to the insatiable sea."

The river's stories are endless. Neither my passengers, nor I, nor other geologists will ever hear them all. There will always be new explanations of the Canyon's history, new interpretations of its various layers. What is essential is that we have been there to listen.

Summary

Erosion—along with weathering, gravity, and transport—has carved the Grand Canyon. All of the landforms that we see in the Canyon are the result of erosion acting upon particular rock types and particular geologic structures. The Colorado River has been actively eroding the Grand Canyon for as long as six million years.

There is a nameless canyon just below Elves Chasm. I'd always floated by it on river trips in the past. But last summer I finally succeeded in camping at its mouth. Since it was not my night to cook, I quickly unloaded my boat and hot-footed it for the canyon's higher ground.

The climb up-canyon was not long, but the route led through a continuous maze of house-sized boulders. After an hour I was at the end of the line, staring up at a 400-foot dry waterfall in the Redwall Limestone. At eye-level the Muav was silver gray; its bedding planes had been worn

to a series of great steps, forming an amphitheater of unlikely proportions. I took out my recorder and began to play. The acoustics were delightful. I played "Four Strong Winds." It is the only song I know. Finishing, I listened to the canyon.

The silence was deafening, so strong that I knew it would absorb all the music I could ever make. I watched the walls and waited.

Suddenly, behind me, a cobble hurtled over the lip of the waterfall. It rattled a little as it lept from the Redwall, tumbled silently, endlessly through the air, and exploded when it crashed into the Muav. Fragments clattered to a resting spot

A VIEW FROM THE BOTTOM

nearby. The silence, momentarily pushed back, quickly returned to engulf even this sound.

This canyon—and the entire Grand Canyon, I realized—is one of the quietest places on earth, and yet it is also one of the most geologically active. Erosion is clawing at the walls of the Canyon at a rate geometrically related to their height. But relative to our lifespans, the Canyon is barely changing at all.

It was not insignificance that I felt as I scrambled back down to dinner. It was, instead, a sense of perspective, a sense of broad-shouldered humility. I felt appropriately small crawling over those creek bottom boulders; I felt appropriately short-lived in the face of the Canyon's lifespan measured in millions of years. I was late for dinner.

An unnamed waterfall in the Muav Limestone.

A Brief Two Billion Year History of the Grand Canyon

Silts, mud and volcanic material are deposited in a coastal environment.	2 billion years ago
Mountain-building occurs; sedimentary and igneous rocks are metamorphosed into the Vishnu Schist.	1.7 billion years ago
Molten rock, squeezed into the Vishnu, cools to become the Zoroaster Granite.	1.2-1.7 billion years ago
Uplift and erosion of the Vishnu Schist and Zoroaster Granite.	
Deposition of the Grand Canyon Supergroup.	0.8-1.2 billion years ago
Block-faulted mountain ranges form, tilting the Grand Canyon Supergroup. Higher sections are eroded away.	
Deposition of Paleozoic sedimentary strata.	0.2-0.5 billion years ago
Mesozoic strata are deposited, but most evidence of their local existence has been eroded from the Grand Canyon region.	0.1-0.2 billion years ago
The Laramide Orogeny uplifts the Colorado Plateau; a compressive event.	65 million years ago
Miocene disturbances create faults along the western Colorado Plateau, further uplifting the Plateau; a tensional event.	20 million years ago
The segments of the Colorado River connect to begin carving the Grand Canyon's configuration.	6 million years ago
Basalts and volcanic ash flow into the western Grand Canyon.	1 million years ago

Bibliography

Geology

Baars, Donald. 1972. *Red Rock Country*. New York: Doubleday Publishing Company.

Breed, William, ed. 1974. *Geology of the Grand Canyon*. Flagstaff, Arizona: Museum of Northern Arizona.

Huntoon, Billingsley, Breed, et al. 1976. *A Geologic Map of the Grand Canyon*. Flagstaff, Arizona: Museum of Northern Arizona and Grand Canyon, Arizona: Grand Canyon Natural History Association.

McKee, Edwin D. 1966. *Ancient Landscapes of the Grand Canyon Region*. Flagstaff, Arizona: Northland Press.

Rahm, David A. 1974. *Reading the Rocks: A Guide to the Geologic Secrets of the Canyons, Mesas, and Buttes of the American Southwest*. San Francisco: Sierra Club.

Stokes, William Lee. 1973. *Scenes of the Plateau Lands and How They Came To Be*. Salt Lake City: Publishers Press.

Background

Babbitt, Bruce. 1978. *Grand Canyon Anthology*. Flagstaff, Arizona: Northland Press.

Crampton, Charles Gregory. 1972. *Land of Living Rock*. New York: Knopf.

Dutton, Clarence E. 1882. *Tertiary History of the Grand Canyon District*. Washington, D.C.: U.S. Geological Survey. Reprinted by Peregrine Smith, 1977, Salt Lake City.

Powell, John Wesley. 1895. *Exploration of the Colorado River and Its Canyons*. Washington, D.C.: Smithsonian Institution. Reprinted by Dover Press, 1961, New York.

Stone, Julius F. 1932. *Canyon Country*. New York: G.P. Putnam's Sons.